# John Logie Baird

## A Pictorial Record of Television Development

### 1924–1938

Foreword by Dr. Graham Winbolt

Introduction by R. W. Burns

KELLY PUBLICATIONS
2001

This edition is limited to 500 copies

First published by
KELLY PUBLICATIONS
6 Redlands, Tiverton, Devon EX16 4DH UK
2001

**ISBN 1-903053-10-2**

# CONTENTS

## NB

The captions in the original album are on typewritten pasted slips, and contain minor inconsistencies and idiosyncrasies. In general these have been left unchanged but, where necessary, the publishers have made minimal corrections and additions in the interest of accuracy. The publisher's additions are included in square brackets.

# Foreword

In 1989 I was invited to view the historical collection of artefacts and documents put together by the Plessey Company in their headquarters at Vicarage Lane, Ilford. This visit had been arranged by Michael Aplin and John McGowan, the former being the then Exhibitions Executive, and the latter the Strategy Director.

Plessey was then in the process of a successful acquisition bid by G.E.C. & Siemens, and it was an emotional experience touring the deserted offices and factory which had been so active in the past.

As a result of this visit I was presented with the residual museum artefacts and paperwork with a view to their preservation. Amongst the books was a rather worn old bound volume with the words "A PICTORIAL RECORD OF TELEVISION DEVELOPMENT" printed in gold leaf upon its badly frayed spine, and inside was an extraordinary collection of photographs illustrating the early experimental television work of John Logie Baird.

Baird had done some of his work in a hut on the flat roof of one of the buildings on the Plessey site, and the company had also been responsible for the manufacture of kits for the original Baird Televisor in 1929. I think the price of a full kit was 25 guineas - quite a lot of money in those days when a copy of *Radio Times* cost two old pence (less than 1p).

I treasured the album for ten years or so, and then, after discussions with Len Kelly, the leading specialist in publications on radio and television who had just started a publishing company together with his wife, it was decided to "liberate" these important photographs to the general public in facsimile form.

The publication of this album coincides with Professor Russell Burns' scholarly biography of Baird, and also with Ralph Barrett's lecture on Baird's life and work to the Royal Institution on 31 January, 2001.

It is hoped that the books will complement each other, and show that Baird was truly a brilliant developer of "seeing by wireless". Without the applied skills of so many scientists during World War II and since, life as we know it today would be impossible. Modern radio, television and radar systems enable us not only to know what is going on in our world, but to travel to all areas of an increasingly shrinking planet. Television broadcasting in particular has had a profound influence on society, mostly for good, and we owe Baird a special debt of thanks for this.

Finally I would like to thank Russell Burns most sincerely for writing the Introduction to this book.

Dr. Graham Winbolt
December 2000

# Introduction

Baird's life and work from the spring of 1923 until the winter of 1925/1926 is unique in the annals of British 20th century electrical engineering. During the early part of 1923, while living and managing a successful business in London, Baird became so ill that his doctor urged him to leave the unhealthy atmosphere of the capital and convalesce on the south coast. Baird decided to stay with a friend from his school days who was residing in Hastings. As his health improved Baird gave some thought to his future career. He considered developing razor blades made from glass, and boots with pneumatic inner soles; finally he resolved to seek a solution to the problem of "seeing by electricity". It was an extraordinary decision to take but Baird embarked on his future life's work with enthusiasm.

It was extraordinary because here was a 35 year-old man with no extended experience of research and development work, no workshop or laboratory facilities, no scientific apparatus of any sort, no employment and no external source of funding, no access to acknowledged expertise or experience, and only one friend in Hastings to give encouragement, seeking in a small room in a suburban house the solution to a problem which had defied the efforts of inventors and scientists around the world for approximately 50 years, and which Dr C V Drysdale, the Superintendent of the British Admiralty Research Laboratory, at the National Physical Laboratory, described in 1926 as "extremely difficult". That Baird succeeded, on 26 January 1926, in demonstrating, for the first time anywhere, even a crude form of television is to his everlasting credit. It was a most remarkable and outstanding achievement which has not been paralleled since. Significantly, ARL was not able to emulate Baird's 1926 achievement even though Dr R T Beatty of ARL had been given a demonstration by Baird.

In the opinion of T H Bridgewater, who worked for Baird from 1928 and who retired as Chief Engineer, BBC Television: "This achievement can never be taken from him. It must not be forgotten and should be glorified for the prodigious pioneer effort it represented - in the same way as other great pioneer feats of British exploration and invention have been honored through the centuries."

Baird now faced a difficulty. His invention had no immediate application to warfare or to safety - unlike Marconi's invention - and he received no patronage from the one body, the British Broadcasting Corporation, which could assist him. He urgently needed funds to progress his ideas, but there seemed to be just one solution to the problem of funding. Baird chose to follow the strategy which had brought financial success to Marconi. Companies were established and the public was invited to purchase shares. However, the public had to be provided with some indication that investment in the companies was beneficial. Hence demonstrations of new applications of Baird's basic system were given. Furthermore, Baird had to acquire a patent holding to enable these companies to have some commercial bargaining power, and, hopefully, be the source of profits from royalties and licensing fees.

From 1926 until circa 1932, Baird succeeded in showing the principles of operation of spotlight scanning, noctovision, phonovision, long distance television, colour television, stereoscopic television, daylight television, zone television, two-way television, and large screen television. His endeavours were, *mutatis mutandis*, wholly consonant with those being undertaken contemporaneously by the mighty and prestigious Bell Telephone Laboratories and deserve much praise.

Baird's pre-war investigations, at his own expense, and in a private laboratory attached to his home, of cinema television (which prior to the outbreak of the Second World War seemed to have a bright future), and his war-time development, under very arduous conditions, of medium and high definition colour television and stereoscopic television are especially commendable. These latter developments were without equal in the United Kingdom at that time.

Looking back over the achievements of Baird it is apparent that he was inspired by the work of Marconi. Baird once said of him: "Although the invention of no single device of fundamental importance can be attributed to Marconi it was he who ventured forth like Christopher Columbus and forced upon the attention of the world the existence of a new means of communication". Baird did the same for British television.

Neither Marconi nor Baird was an original scientist; rather they were tenacious experimenters who, with great resolution and initially single-handedly, sought to demonstrate narrow band wireless broadcasting and narrow band television broadcasting respectively, and to advance the possibilities which their work appeared to suggest.

Sir Noel Ashbridge, a former Chief Engineer of the BBC, once remarked: "There is no doubt that the experiments of J L Baird accelerated serious and urgent consideration of the practical possibilities of a public television service." He should be remembered, as Bridgewater has written, as "a kindly, courteous, sensitive and brave man - a man with a passionate faith in television's destiny in the service of mankind..."

R. W. Burns
December 2000

Mr. John L. Baird carrying out early Television experiments in March 1925

Early Baird Television experiments

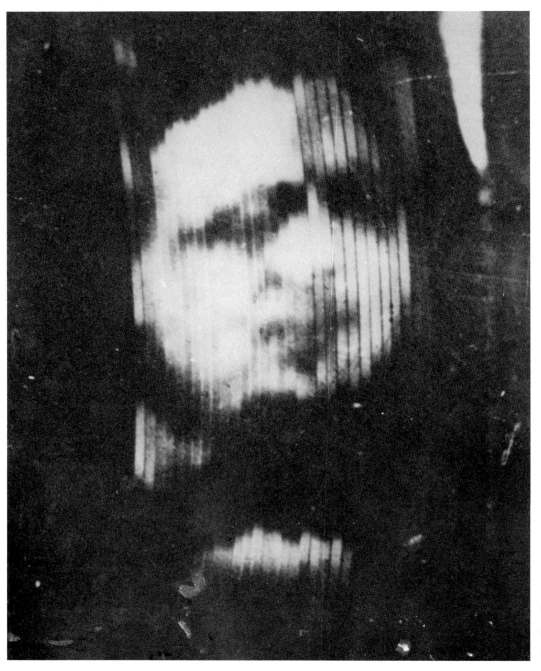

The first television photograph ever taken. This is an actual unretouched photo of the image as it appeared on the screen of the first "Televisor"

3

Baird with his very crude transmitter equipment in 1925

Some of the early Baird transmitter equipment

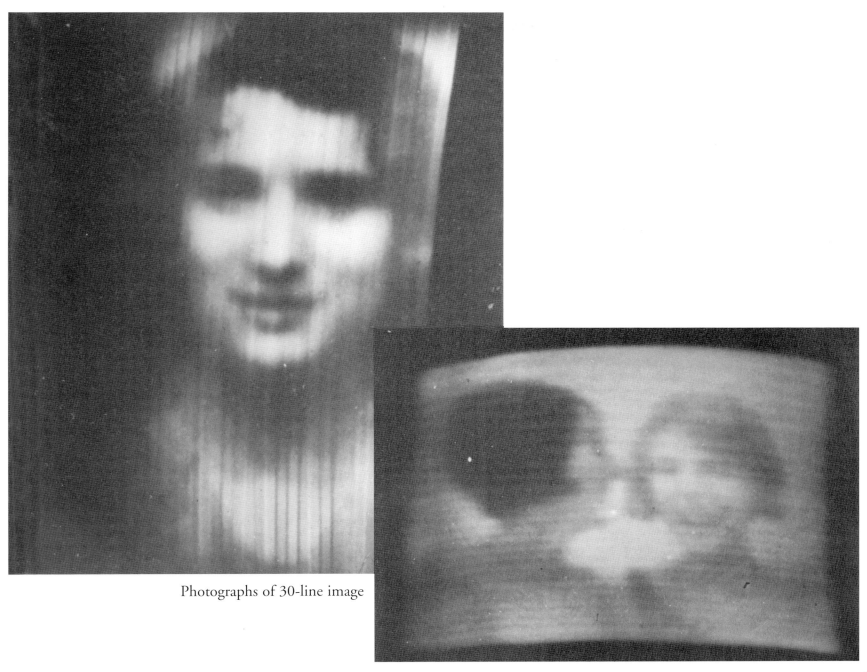

Photographs of 30-line image

6

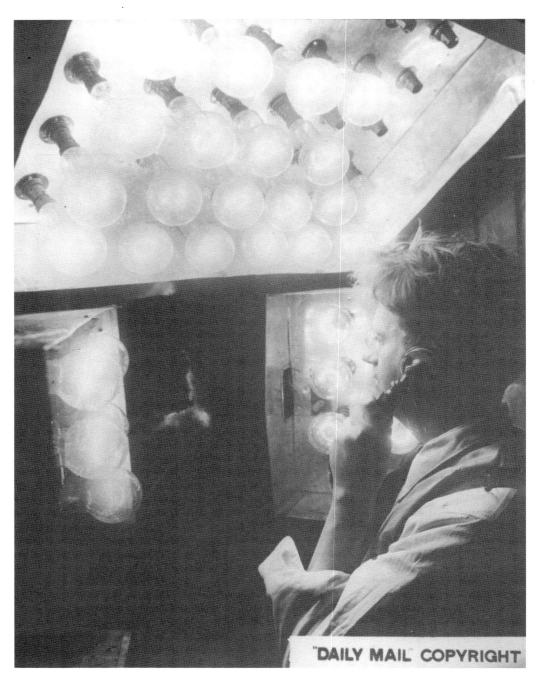

"DAILY MAIL" COPYRIGHT

1927 - Early spotlight transmitter. J. L. Baird at Transmitter for London to Glasgow experiment

7

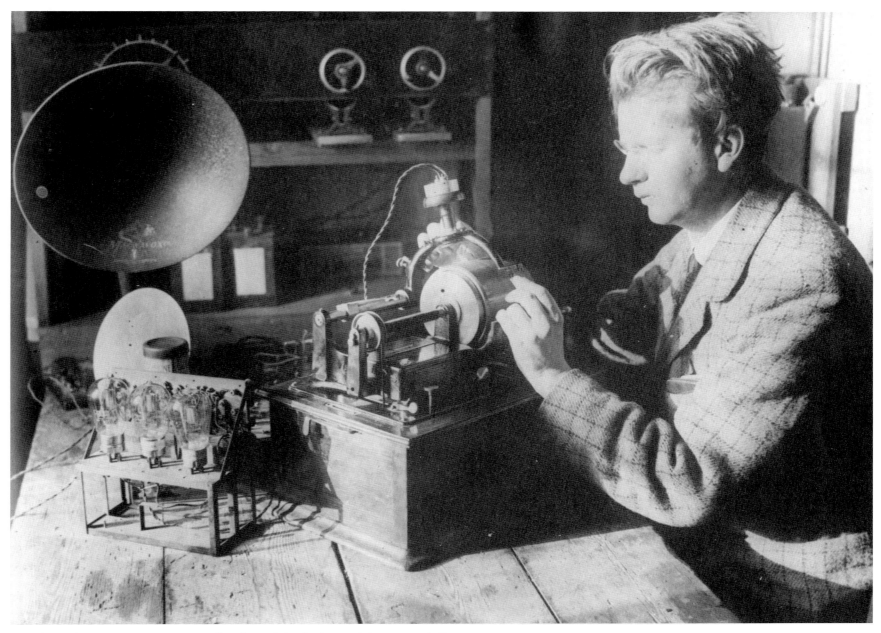

Baird conducting an experiment for the original phonovision record, January 1927

Baird with the original transmitter used for the first transmission of signals across the Atlantic in 1928

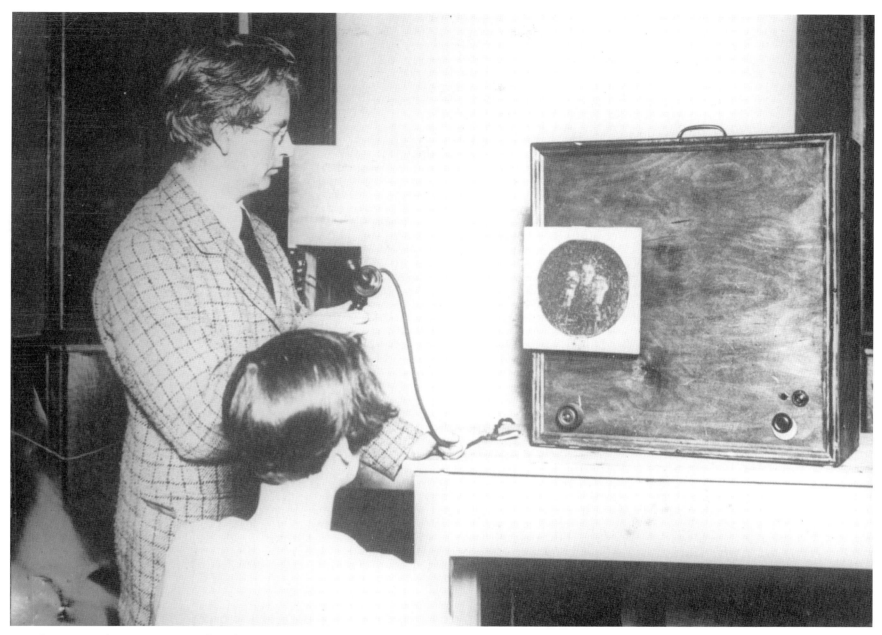

Baird giving a demonstration of 30-line reception

Arthur Prince and his doll being televised in 1928

11

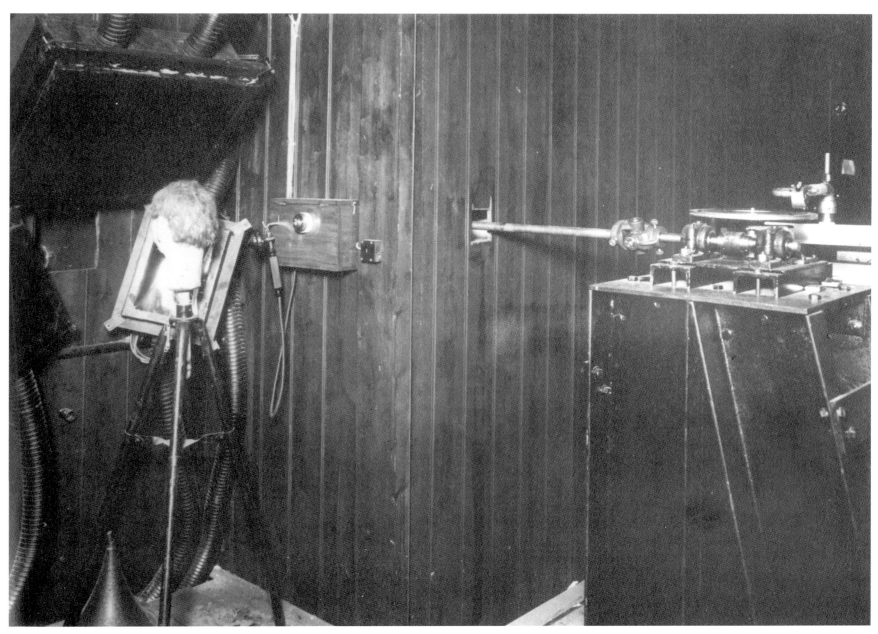

A dummy's head being used as the subject for producing a phonovision record

Baird giving a demonstration of 30-line reception at Selfridges

One of the original
photo-electric cells used
for the Baird 30-line
spotlight transmitter

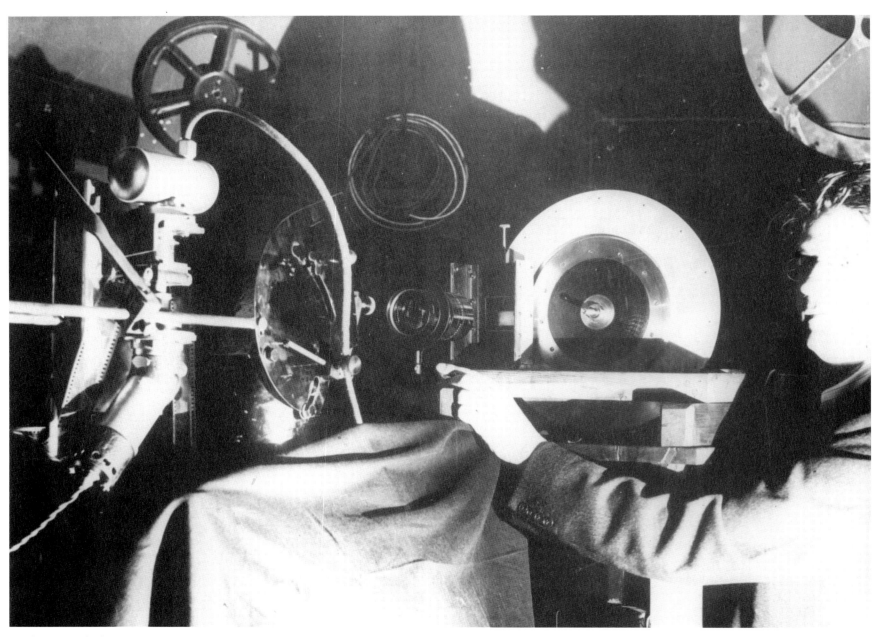

30-line spotlight transmission and control room used at Long Acre studio in 1928/30

15

The first demonstration of television [to be received in Bradford from the BBC in London]    [Seated: H. J. Barton Chapple.  Extreme left: Sydney A. Moseley]

16

Dénes von Mihály's original
30-line receiver using periscope
observation (about 1930)

American 30-line telecine
transmission        1930

18

a. b.
2 ft home receiver
optical chassis

c. A Baird kit of parts to
build a complete receiver

Plessey Baird disc machine used on the first Television reception

Plessey receiver used in conjunction with the disc Television receiver being demonstrated at the Baird studios in Long Acre

21

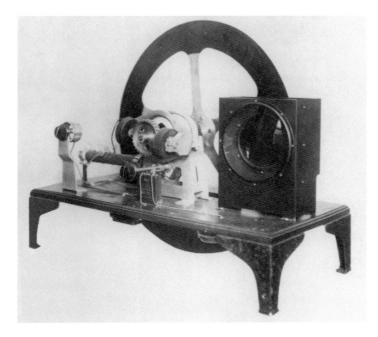

Baird television receiver made at the Plessey Co. 1931/2. Showing front back and interior also phonic wheel synchronising mechanism and motor drive

A German Television receiver with mirror drum

23

Baird spotlight transmitter
as supplied to B.B.C.

Early disc Television receiver
used by Germans in 1931

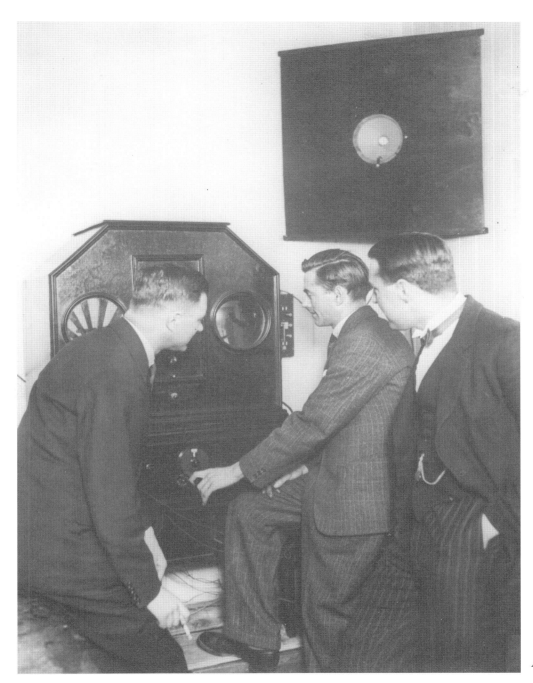

An early 30-line receiver

The first tele-talkie apparatus used. Transmitting talking film on 30 lines

30-line television transmitter as supplied to B.B.C. for installation at Broadcasting House

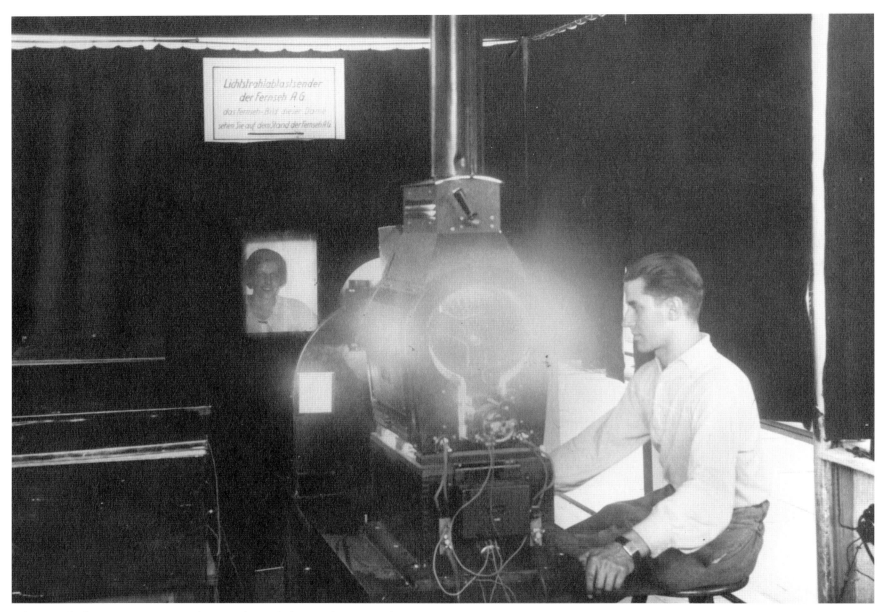

The spotlight transmitter used at Berlin exhibition by Fernseh in 1931

29

Receiving a picture in mid Atlantic on the BERENGARIA using a 30-line Baird receiver

The spotlight transmitter used by Fernseh in Berlin

The ultra short-wave transmission aerial used on the top of the south tower of Crystal Palace worked on 6/7 meters

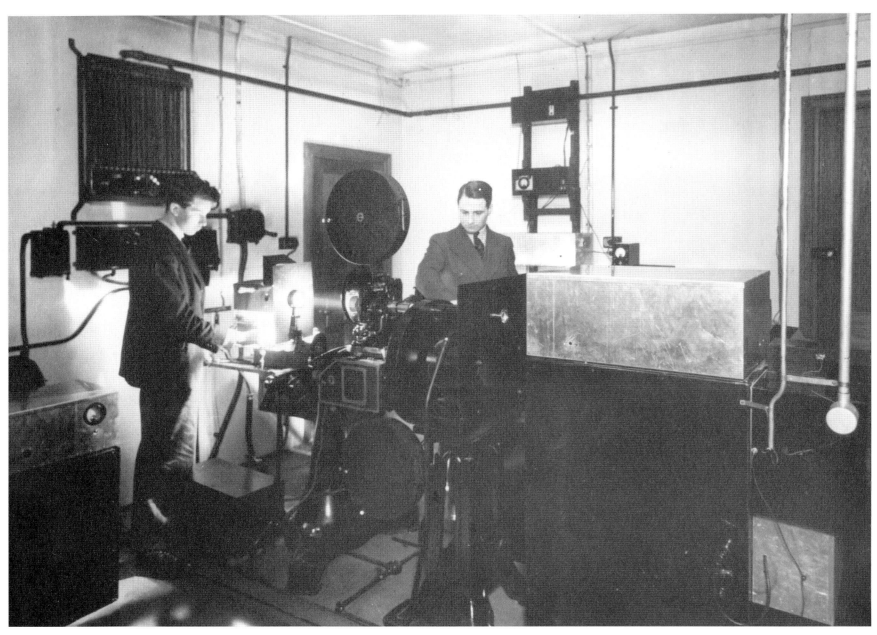

240-line Telecine transmitter as used originally at Crystal Palace 1934-35

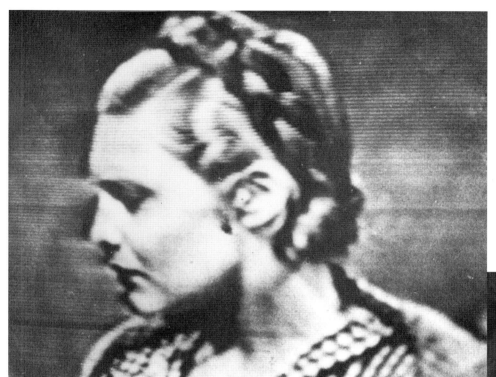

Pictures of Madeline Carroll as she appeared in the film "I was a Spy" which was the film used by the Baird Company for original transmission tests from Crystal Palace

Early television experiments by designers L. H. Bedford and O. S. Puckle

Baird at the un-veiling of the plaque by the Mayor of Hastings, where Baird first demonstrated Television in 1924

30-line mirror drum receiver 1935. Using a projection Neon Tube

37

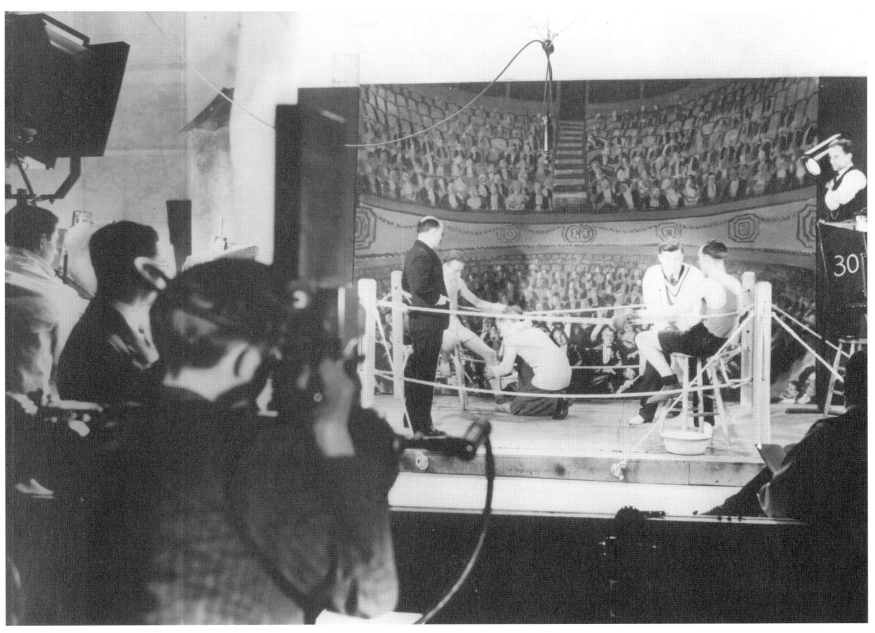

Televising a boxing match using the Baird intermittent film transmitter at Crystal Palace

Late Sir Ambrose Fleming with John Baird on the set at Crystal Palace

39

The first regular television programme from Alexandra Palace [Douglas Birkinshaw, first BBC Television Chief Engineer, behind the camera]

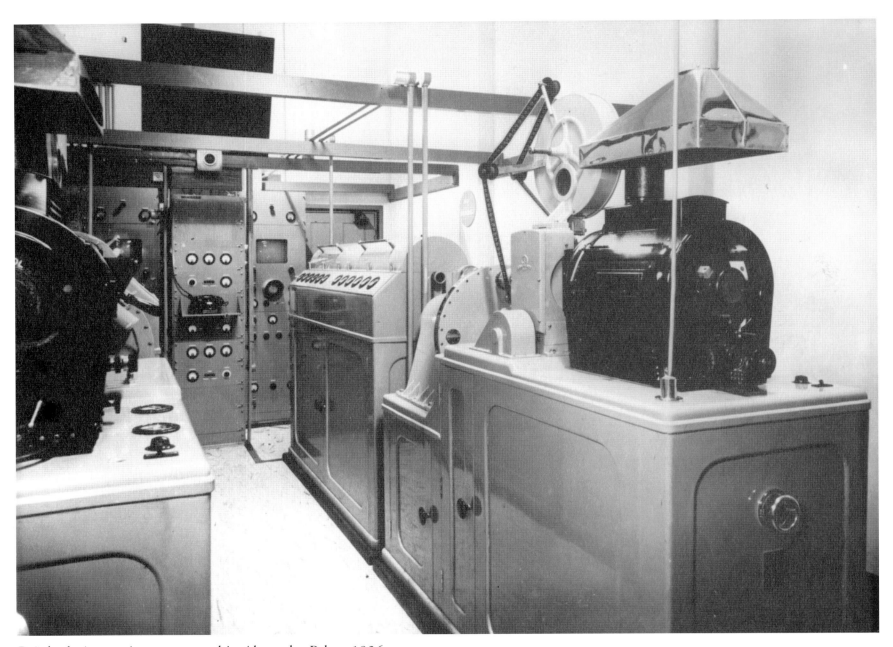

Baird telecine equipment as used in Alexandra Palace 1936

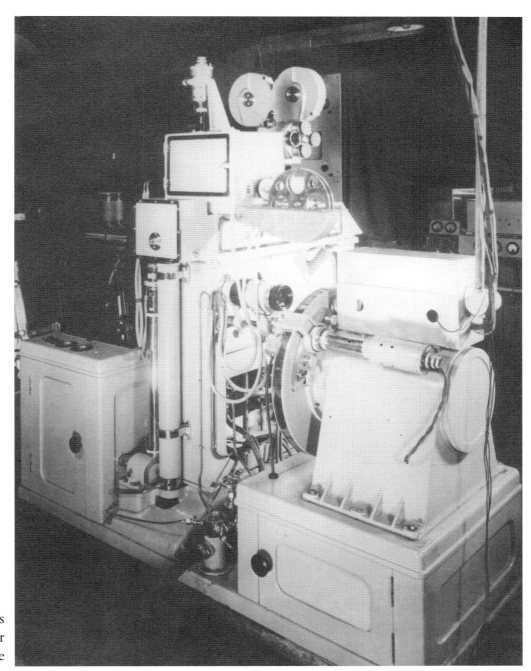

Baird film equipment as supplied to B.B.C. for installation at Alexandra Palace

Baird film equipment as supplied to B.B.C. for installation at Alexandra Palace

Intermediate Film transmitter car, delivered to the German Reich-Rundfun-Ges and subsequently used for the transmission of exterior scenes by the German Post Office. The transmission of Outdoor scenes by Television

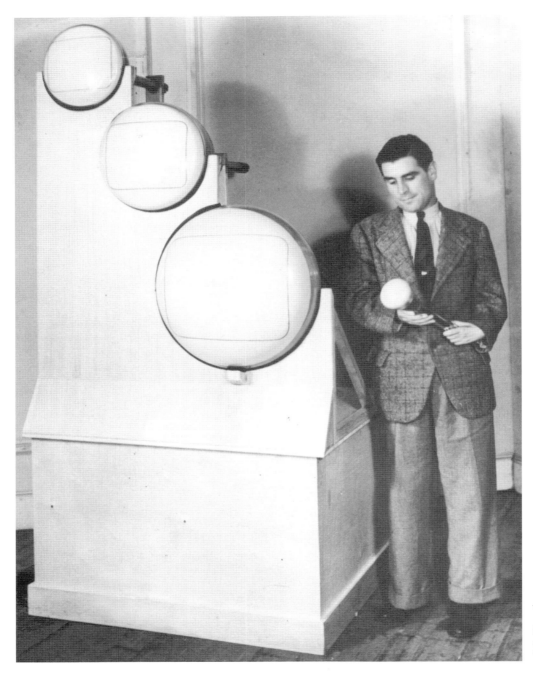

William Taynton the first boy [actually the first *person*] to be televised by Baird in 1925, holding various sized cathode ray tubes

The first projection picture showing an 8'x 6' screen in the Tatler Cinema 1938

Projection television receiver installed in the Marble Arch Cinema for the receiving of special television programme prior to the outbreak of 1939 war. One projection tube also shown

Baird television van for outside broadcasts

Baird television van
for outside broadcasts

The original caravan trailer equipment which housed the 2,100 lamp screen for demonstration to the public on the stage of the Coliseum in 1930

50

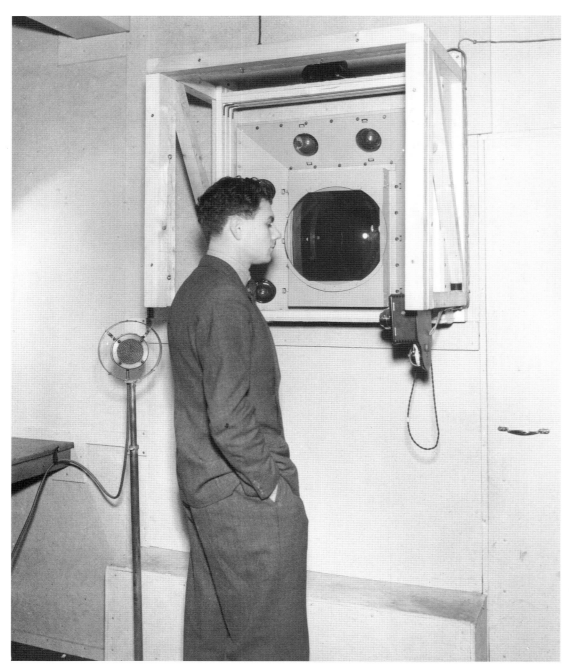

The photo-electric cells used for the multi-zone spotlight transmitter by Baird

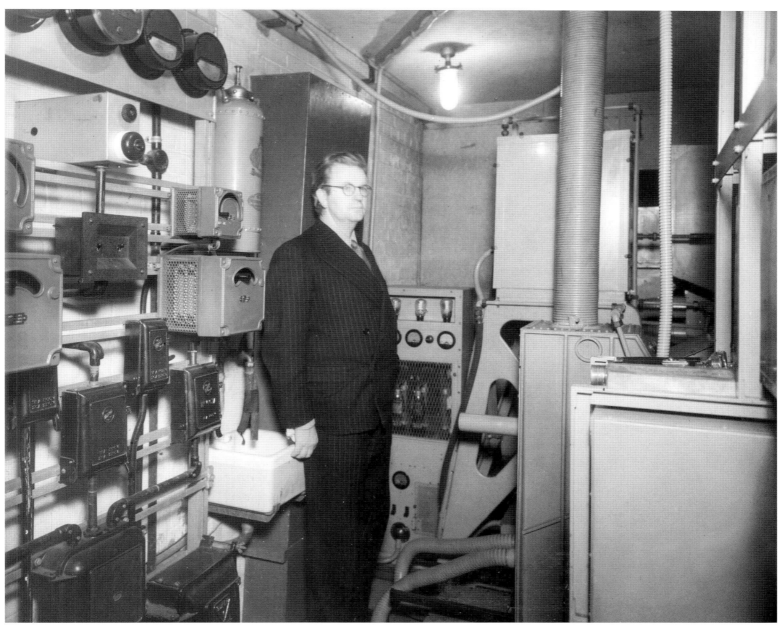

Mr. Baird inside the room housing the multi-zone receiving equipment when it was demonstrated at a London cinema

The large mirror drum, housed in the caravan at Epsom in 1932, with its stroboscoptic sectors for accurate speed adjustment. This was part of the equipment used for a 3-zone televising of the Derby direct to the Metropole Cinema, London

The original cathode ray tube large screen projection equipment used in the Tatler Cinema, London, on the occasion of the Trooping the Colour 1938

[Same photograph as in Plate 46]

One of the Baird laboratories in Long Acre showing the rear view of a Plessey television receiver, while on the bench on the left is one of the original Plessey receiver amplifiers developed for use in conjunction with the disc receiver

The radio transmitter used for sending the first television pictures across the Atlantic in 1927 was accommodated in this house

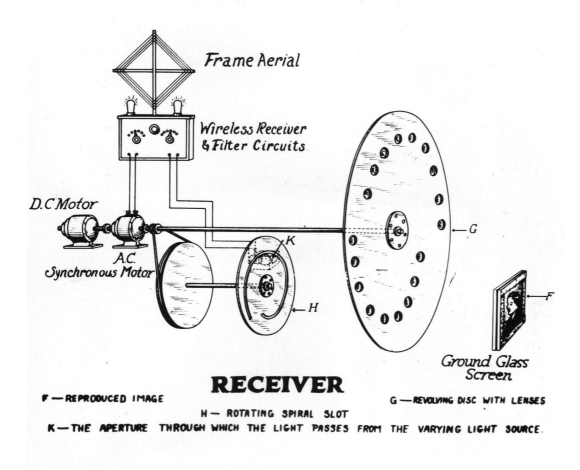

**Frame Aerial**

**Wireless Receiver & Filter Circuits**

**D.C. Motor**

**A.C. Synchronous Motor**

*K*

*H*

*G*

*F*

**Ground Glass Screen**

# RECEIVER

F — REPRODUCED IMAGE

G — REVOLVING DISC WITH LENSES

H — ROTATING SPIRAL SLOT

K — THE APERTURE THROUGH WHICH THE LIGHT PASSES FROM THE VARYING LIGHT SOURCE.

A pictorial representation of one of Mr. Baird's original television receivers

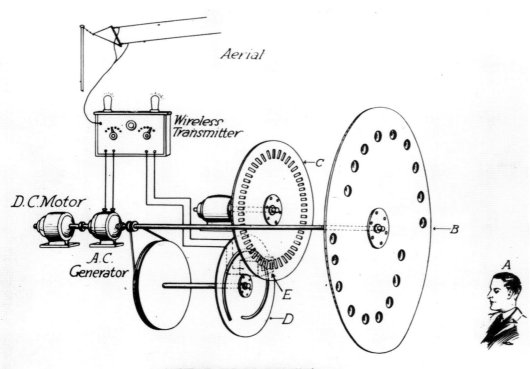

Aerial

Wireless
Transmitter

D.C.Motor

A.C.
Generator

C

B

E

D

A

# TRANSMIT·TER

**A –** THE OBJECT TO BE TRANSMITTED     **B –** A REVOLVING DISC WITH LENSES
**C –** A SLOTTED DISC REVOLVING AT HIGH SPEED     **D –** A ROTATING SPIRAL SLOT
**E –** THE APERTURE THROUGH WHICH THE LIGHT PASSES TO THE LIGHT SENSITIVE CELL

A pictorial representation of one of Mr.
Baird's original television transmitters

The first model of the mirror drum home projection
30-line television receiver as developed in 1930

The commutator used for the Baird
lamp screen was a massive affair,
while the wiring to all the bars was
a very complicated arrangement

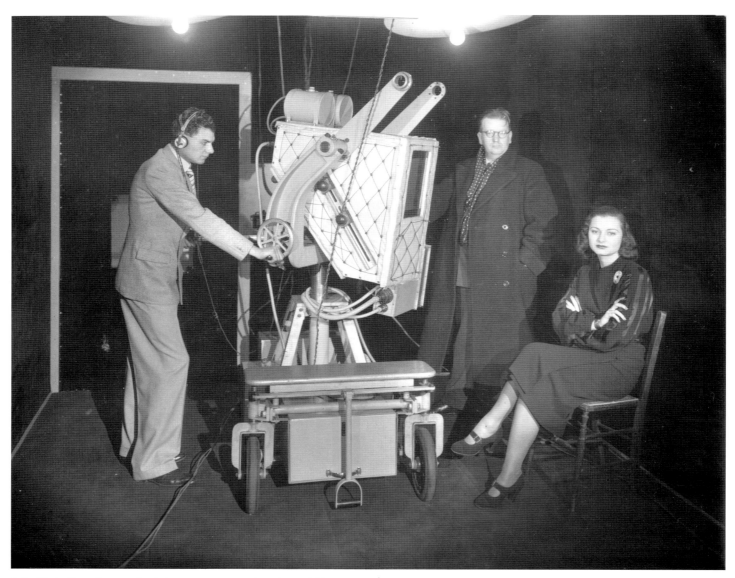

Mr. Baird's colour television camera as developed for
showing high definition colour television pictures

Some of the various types of neon tubes used by the
Baird Company in the early 30-line television days

One of the original large projection cathode ray tubes used by Baird for his original direct projection of television pictures on to a cinema screen

An interior view of a portable
30-line television receiver as used
for the provincial experiments

An external view of a portable 30-line television receiver as used for the first successsful demonstration of television in the provinces in 1929

The printed script passing in front of
the televison transmitting apparatus
for conversion into a television signal

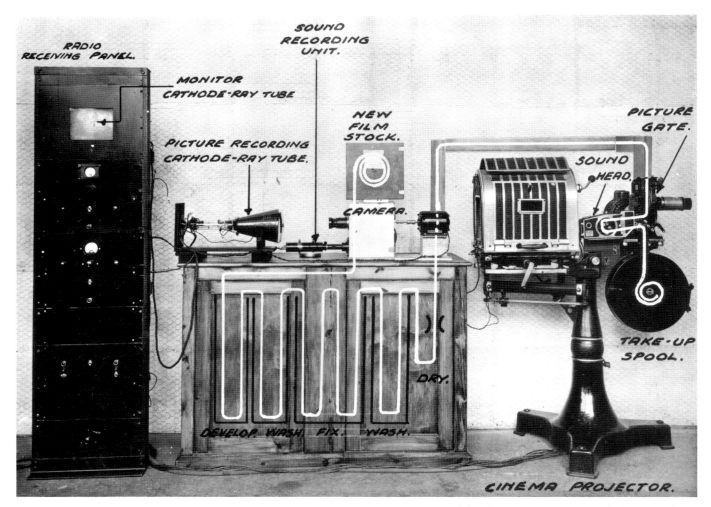

RADIO
RECEIVING PANEL.

SOUND
RECORDING
UNIT.

MONITOR
CATHODE-RAY TUBE

PICTURE RECORDING
CATHODE-RAY TUBE.

NEW
FILM
STOCK.

PICTURE
GATE.

CAMERA.

SOUND
HEAD.

DEVELOP. WASH. FIX. WASH.

DRY.

TAKE-UP
SPOOL.

CINEMA PROJECTOR.

A prototype of the first equipment used for recording television pictures on a film which was subsequently developed, washed and dried, and passed into a film projector for projecting on to a cinema screen

a. The original 30-line disc television transmitter in the headquarters of the German Post Office of Berlin, and used for carrying out experimental transmissions from the Witzleben Tower in 1929

b. Showing the disc television receiver as made by the Plessey Company in use in a laboratory in Austria, where television pictures were received in the early 30-line days

The large screen used by Baird
for showing high definition
colour television pictures

69

George Robey as "The Blushing Bride" was the first film ever to be publicly televised by the Baird 30-line equipment

An example of real modern home "entertainment" as illustrated by a Baird de luxe television receiver built before the war. It incorporated a 15" cathode ray tube, all-wave radio, automatic record-changing gramophone and a cocktail cabinet

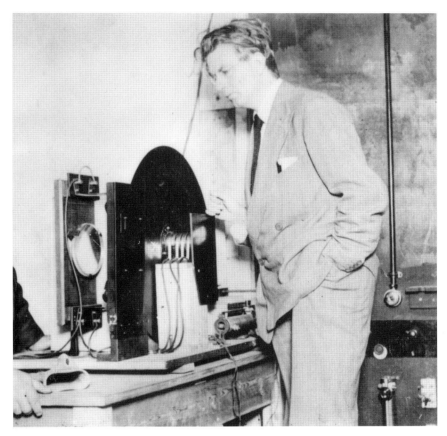

Mr. Baird, with his original three-colour
television disc receiving equipment